Motion of Charged Particles in the Earth's Magnetic Field

DOCUMENTS ON MODERN PHYSICS

Edited by

ELLIOTT W. MONTROLL, Institute of Defense Analyses
GEORGE H. VINEYARD, Brookhaven National Laboratory

A. H. COTTRELL, Theory of Crystal Dislocations
A. ABRAGAM, L'Effet Mössbauer
A. B. PIPPARD, The Dynamics of Conduction Electrons
R. GALLET, Whistlers and VLF Emissions
K. G. BUDDEN, Lectures on Magnetoinic Theory
S. CHAPMAN, Solar Plasma, Geomagnetism and Aurora
R. H. DICKE, The Theoretical Significance of Experimental Relativity
J. A. WHEELER, Geometrodynamics and the Issue of the Final State
BRYCE S. DEWITT, Dynamical Theory of Groups and Fields
JOHN G. KIRKWOOD, Selected Topics in Statistical Mechanics
JOHN G. KIRKWOOD, Macromolecules
JOHN G. KIRKWOOD, Theory of Liquids
JOHN G. KIRKWOOD, Theory of Solutions
JOHN G. KIRKWOOD, Proteins
JOHN G. KIRKWOOD, Quantum Statistics and Cooperative Phenomena
JOHN G. KIRKWOOD, Shock Waves
JOHN G. KIRKWOOD, Transport Processes
JOHN G. KIRKWOOD, Dielectrics—Intermolecular Forces—Optical Rotation

Additional volumes in preparation

Motion of Charged Particles in the Earth's Magnetic Field

JOSEPH W. CHAMBERLAIN

Yerkes Observatory and Department of Geophysical Sciences, University of Chicago

GORDON AND BREACH

Science Publishers

New York • London

150 Fifth Avenue, New York, New York 10011

Library of Congress Catalog Card Number 64-25800

These lectures were presented at the Les Houches Summer School of Theoretical Physics, and published in the proceedings volume, Geophysics, the Earth's Environment, C. De Witt et al, Eds., Gordon & Breach, 1963.

The author has corrected and amended the lectures for this edition.

PRINTED IN UNITED STATES OF AMERICA

EDITOR'S PREFACE

Seventy years ago when the fraternity of physicists was smaller than the audience at a weekly physics colloquium in a major university, a J. Willard Gibbs could, after ten years of thought, summarize his ideas on a subject in a few monumental papers or in a classic treatise. His competition did not intimidate him into a muddled correspondence with his favorite editor nor did it occur to his colleagues that their own progress was retarded by his leisurely publication schedule.

Today the dramatic phase of a new branch of physics spans less than a decade and subsides before the definitive treatise is published. Moreover, modern physics is an extremely interconnected discipline and the busy practitioner of one of its branches must be kept aware of breakthroughs in other areas. An expository literature which is clear and timely is needed to relieve him of the burden of wading through tentative and hastily written papers scattered in many journals.

To this end we have undertaken the editing of a new series, entitled *Documents on Modern Physics,* which will make available selected reviews, lecture notes, conference proceedings, and important collections of papers in branches of physics of special current interest. Complete coverage of a field will not be a primary aim. Rather, we will emphasize readability, speed of publication, and importance to students and research workers. The books will appear in low-cost paper-covered editions, as well as in cloth covers. The scope will be broad, the style informal.

From time to time, older branches of physics come alive again, and forgotten writings acquire relevance to recent developments. We expect to make a number of such works available by including them in this series along with new works.

ELLIOTT W. MONTROLL
GEORGE H. VINEYARD

PREFACE

This brief introduction to one area of plasma physics is directed primarily to geophysicists, but the geophysical problems themselves are not treated here. I have dealt mainly with the orbit theory of particles in a magnetic field, with particular emphasis on the geomagnetic field.

On the whole, cosmical electrodynamics (a rough astronomical equivalent to the more mundane "plasma physics") has become a highly developed and complex topic, largely through the contributions of physicists primarily concerned with laboratory plasmas. Undoubtedly plasma physics will continue to be exploited in the geophysical realms of aurorae, geomagnetism, magnetosphere, ionosphere, etc. But the process is slowed by the difficulty geophysicists often have in acquiring a working knowledge of plasma physics.

I hope, therefore, that the review will be particularly useful to graduate students and others who may wish to work through the mathematics to gain some physical insight into the details, albeit the elementary ones, of the topic. The article includes a simplified treatment of Northrup and Teller's demonstration of the second adiabatic invariant, which I especially hope will be of some use.

J. W. C.

Tucson, Arizona

July 1964

CONTENTS

The theoretical description of the motions of charged particles in magnetic fields has proceeded historically from the exact theory for approximate physical models to approximate methods for more accurate physical situations. The first description of the geomagnetic control of charged particles was Poincaré's (1896) solution of the equation of motion of a particle in the field of a monopole, which was inspired by Birkeland's terrella experiments on the aurora. Störmer (1907, 1911, 1955) then undertook the solution of the dipole problem, which can be partially solved analytically, but which requires a numerical solution for the third integral of the equation of motion. This problem is examined in Section 1.

Because of difficulties involved in the exact solution of the dipole problem, Alfvén (1950) began the development of approximate techniques valid for particles of small momentum. This approach has been further developed and widely used in recent years, not only in the description of trapped particles in the geomagnetic field but in laboratory plasmas as well. These matters are discussed in Section 2 where we also allow for the possibility of macroscopic electric fields that affect a particle's motion.

In Section 3 we will review briefly the description of motions of trapped particles with the macroscopic equations of hydromagnetics.

1. The Dipole Problem[1]

Störmer (1907, 1911, 1955) developed the theory for a particle in a dipole magnetic field with an explanation of aurora in mind. While it is now clear that the auroral particles cannot be so simply described, the theory is still of fundamental importance. Being an elegant mathematical characterization of a specific physical problem it is still of some practical importance in cosmic-ray theory.

We assume that there are no electrical interactions—neither electrostatic forces between individual particles nor interactions between their separate currents. We are thus concerned with the motion of an isolated charged particle according to the equation of motion

$$m\frac{d\mathbf{v}}{dt} = \pm \frac{e}{c}\mathbf{v} \times \mathbf{B}, \tag{1}$$

where v is the velocity, m the mass, and e the absolute value of the charge on the particle. The upper sign is applied to positive ions, the lower one to electrons.

[1] Based on a review by Chamberlain (1958), which also contains a summary of Störmer's application of the theory to aurora (courtesy of Academic Press). Gaussian units are employed throughout.

If there is no electric field, the energy of the particle remains constant. Multiplying equation (1) by $\mathbf{v}$ we have

$$m\mathbf{v}\cdot\frac{d\mathbf{v}}{dt} = mv\frac{dv}{dt} = \frac{d}{dt}\left(\frac{m}{2}v^2\right) = 0, \tag{2}$$

where $v = |\mathbf{v}|$. Thus, as the acceleration is always perpendicular to the velocity, the scalar speed does not change.

1.1 The Dipole Field

The potential of a dipole field may be readily derived by adding (algebraically) the scalar potentials of the two single poles, each of which has an inverse-square field. Thus we find the dipole potential Ω_P at a point P is

$$\Omega_P = \frac{\mathbf{M}\cdot\mathbf{r}}{r^3} = -\mathbf{M}\cdot\nabla\left(\frac{1}{r}\right), \tag{3}$$

where $\mathbf{r}$ is the vector from the dipole to P, and where the gradient is taken at the point P (that is, the dipole is held fixed). Here $\mathbf{M}$ is the magnetic moment of the dipole. For the earth's field $M = |\mathbf{M}| = 8.1 \times 10^{25}$ gauss cm^3. The magnetic field is then

$$B = -\boldsymbol{\nabla}\Omega = \boldsymbol{\nabla}\left[\mathbf{M}\cdot\boldsymbol{\nabla}\left(\frac{1}{r}\right)\right] = 3(\mathbf{M}\cdot\mathbf{r})\frac{\mathbf{r}}{r^5} - \frac{\mathbf{M}}{r^3}\cdot \tag{4}$$

The earth's dipole field is oriented with the "south" pole of the magnetic dipole in the northern hemisphere (the axis point is near Thule, Greenland). The situation is a bit confusing, because we usually speak of the Thulé pole as the "north magnetic pole". However according to the convention that has been set up, the "north" pole of, say, a compass magnet will orient itself to point northward. Hence it points toward the "south" pole of the geomagnetic field.

Thus we may picture the lines of force of the geomagnetic field as proceeding from south to north outside the earth and from north to south within the "magnet". We shall choose the z-axis along the axis of the dipole and positive toward the north. That is, the dipole moment $\mathbf{M}$ is oriented along the $-z$ direction.

With our axis so chosen, we have in Cartesian coordinates, from equation (4)

$$B = \boldsymbol{\nabla}\frac{Mz}{r^3} = -\frac{3Mxz}{r^5}, -\frac{3Myz}{r^5}, -\frac{M}{r^5}(3z^2 - r^2). \tag{5}$$

From this equation it is readily seen that

$$B = (B_x^2 + B_y^2 + B_z^2)^{\frac{1}{2}} = \frac{M}{r^3}\left(1 + \frac{3z^2}{r^2}\right)^{1/2}. \tag{6}$$

Thus at the equator ($z = 0$) or at the poles ($z = r$) the field decreases with the inverse cube of the distance. At the surface of the earth the field strength is a little over 0.3 gauss at the equator and nearly 0.7 gauss at the poles.

1.2 Integrals of the Equation of Motion

For the dipole field given by equation (5), the equation of motion (1) for a positively charged particle is

$$\frac{d\mathbf{v}}{ds} = \frac{eM}{mvc}\mathbf{v} \times \nabla\left(\frac{z}{r^3}\right), \tag{7}$$

where we use the path length s, defined by $ds = v\,dt$, as the independent variable in place of it.

We adopt as the unit of length the value

$$C_{st} = \left(\frac{Me}{mvc}\right)^{1/2}. \tag{8}$$

This "Störmer unit" will be used through the remainder of this discussion of Störmer's theory, except where otherwise explicitly stated. The equation of motion is then, for a positive particle,

$$\frac{d\mathbf{v}}{ds} = \mathbf{v} \times \nabla\left(\frac{z}{r^3}\right). \tag{9}$$

For a negatively charged particle we may use the same solution and merely reverse the direction of motion about the dipole axis; in other words, we might represent the solution for an electron in the left-handed system of coordinates instead of the right-handed system adopted here.

The first integral may be obtained by taking the scalar product of $\mathbf{v}$ with equation (9), as we did with equation (2). We thus obtain the condition that the kinetic energy remains constant. In cylindrical coordinates we may write the integral of equation (2) as

$$\left(\frac{dR}{ds}\right)^2 + R^2\left(\frac{d\varphi}{ds}\right)^2 + \left(\frac{dz}{ds}\right)^2 = 1 \tag{10}$$

where the integration constant is unity by virtue of our use of s instead of t as independent variable. To relate s and t we need to use the unit of length in equation (8); hence we may consider C_{st} (or v) as an integration constant to be fixed by the initial conditions.

A second integral is obtained by integrating one of the scalar equations composing the vector equation (9). In cylindrical coordinates the v_ϕ component of equation (9) is

$$2\frac{dR}{ds}\frac{d\varphi}{ds} + R\frac{d^2\varphi}{ds^2} = \left(\frac{3z^2}{r^5} - \frac{1}{r^3}\right)\frac{dR}{ds} - \frac{3zR}{r^5}\frac{dz}{ds}. \tag{11}$$

Multiplying this equation through by R, we obtain an expression equivalent to

$$\frac{d}{ds}\left(R^2\frac{d\varphi}{ds}\right) = \frac{d}{ds}\left(\frac{R^2}{r^3}\right). \tag{12}$$

This relation is immediately integrable, and we have the angular momentum integral,

$$R^2\frac{d\varphi}{ds} = \frac{R^2}{r^3} + 2\gamma, \tag{13}$$

where 2γ is the integration constant.

1.3 The Equations of Motion for the Meridian Plane

The complete formal solution of the problem requires a third integral. Since this has never been found, it is necessary to carry out integrations numerically.

Many properties of the motions in a dipole field may be obtained, however, from a study of the differential equations combined with partial solutions as given by the kinetic-energy and angular-momentum integrals.

Let

$$Q = 1 - \left(\frac{2\gamma}{R} + \frac{R}{r^3}\right)^2 = \left(\frac{dR}{ds}\right)^2 + \left(\frac{dz}{ds}\right)^2, \tag{14}$$

where the second equality is obtained by eliminating $d\varphi/ds$ between equations (10) and (13).

Now we write the R and z components of equation (9):

$$\frac{d^2R}{ds^2} = R\left(\frac{d\varphi}{ds}\right)^2 + \frac{r^2 - 3z^2}{r^5}R\frac{d\varphi}{ds} \tag{15}$$

and

$$\frac{d^2z}{ds^2} = \frac{3zR^2}{r^5}\frac{d\varphi}{ds}. \tag{16}$$

If we replace $d\varphi/ds$ in these equations by equation (13), and then differentiate equation (14) (where we use the first equality sign), we find that these equations of motion may be expressed as

$$\frac{d^2R}{ds^2} = \frac{1}{2}\frac{\partial Q}{\partial R} \tag{17}$$

and

$$\frac{d^2z}{ds^2} = \frac{1}{2}\frac{\partial Q}{\partial z}. \tag{18}$$

These two equations describe the motion of the particle in its meridian plane. From equation (14) it is apparent that Q is the kinetic energy of motion in this plane. On the other hand, the meridian plane for the particle is itself in motion with a variable angular velocity $d\varphi/ds$ given by equation (13).

1.4 Orbits Lying in the Equatorial Plane

It is instructive to review this two-dimensional problem, which can be solved analytically, before passing to the more general auroral orbits. We assume the particles lie in the equatorial plane and have no z component of velocity. For this problem it is not necessary to use the equations of motions involving Q.

For the second equality in (14), remembering that $r = R$ in the equatorial plane, we have

$$\frac{dR}{ds} = \left[1 - \left(\frac{2\gamma}{R} + \frac{1}{R^2}\right)^2\right]^{1/2}. \tag{19}$$

Equation (13) gives

$$\frac{d\varphi}{ds} = \frac{1}{R^3} + \frac{2\gamma}{R^2}. \tag{20}$$

Combining these two equations to eliminate ds, we have

$$\frac{d\varphi}{dR} = \frac{(2\gamma R + 1)}{R[R^4 - (2\gamma R + 1)^2]^{1/2}}, \tag{21}$$

which can be integrated by elliptic functions and thus represents the formal solution to the problem.

For our present purposes this solution itself is of little interest; it is more instructive from the standpoint of the so-called "forbidden regions," to examine the properties of equation (21). For a real solution, the square root must obviously be real. That is, we must have

$$R^2 > (2\gamma R + 1) > -R^2, \tag{22}$$

where we consider the possibility of either positive or negative values of γ. Hence the allowed values of R, when γ is specified, must obey the inequalities

$$\frac{R^2 - 1}{2R} > \gamma \tag{23}$$

and

$$-\frac{R^2 + 1}{2R} < \gamma. \tag{24}$$

In Fig. 1 are plotted the curves (23) and (24) with equalities replacing the inequalities. These relations between R and γ then imply that a particle can exist only to the left of curve (23) and to the right of curve (24). All other regions on the diagram are forbidden.

In Fig. 1 the orbit of a particle coming from infinity is represented by a straight vertical line, at some value of γ. The particle approaches the $R = 0$ axis (the dipole) until it reaches one of the boundaries between the permitted and inaccessible regions. At this point it is magnetically reflected back again to infinity along the same value of γ.

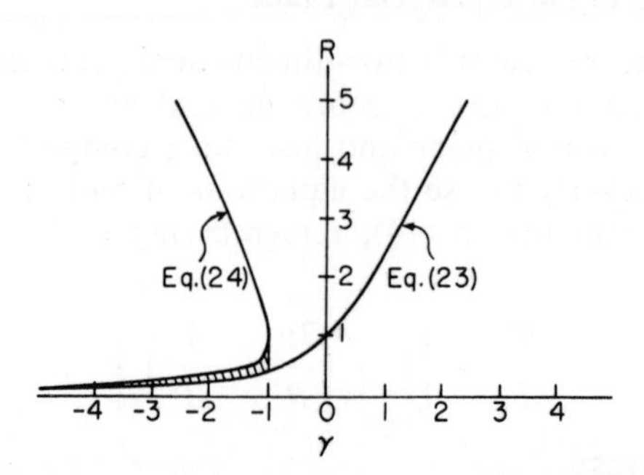

Fig. 1. Permitted regions in the equatorial plane *versus* the integration constant γ for a Störmer particle. Curves plotted from equations (23) and (24) are indicated and explained in detail in the text. The shaded region contains captive orbits.

Toward the left of the figure, there is a shaded region bounded on top by curve (24) and below by curve (23). A particle from infinity cannot penetrate into this region (without some perturbation to its orbit) and a particle once in this region will remain there. These are the "captive orbits". In Fig. 1 these particles move up and down, bouncing between the two curves, which act as barriers to inaccessible regions. These captive orbits must have $\gamma < -1$.

Another interpretation of Fig. 1 is possible. For a given point R in space, γ determines uniquely the direction of motion of the particle. If we let ω be the angle between the tangent to the orbit and the R axis, then the "velocity"[2] in the φ direction is

$$R\frac{d\varphi}{ds} = \sin \omega, \tag{25}$$

since the total velocity is one in our units. Then from equation (20) we immediately find

$$\gamma = \frac{R^2 \sin \omega - 1}{2R}. \tag{26}$$

In Fig. 1, curve (23) corresponds to $\sin \omega = +1$ and (24) to $\sin \omega = -1$. At large distances R the algebraic sign of $\sin \omega$ is the same as the sign of γ. Hence particles with $\gamma < -1$ and those with $\gamma > 0$ move in one sense throughout their trajectories, as seen from the origin. On the other hand, when $-1 < \gamma < 0$, the sign of $\sin \omega$ reverses at some point on the orbit. For those values of γ close to -1,

[2]By "velocity" we shall often mean the derivative of the particle's position with respect to arc length s rather than time t. In this usage the total velocity is always unity.

this reversal takes place quite rapidly and a small loop appears in the orbit (see Fig. 2).

The closest possible approach to the dipole occurs when $\gamma = -1$ and the particle is reflected from curve (23). We thus find $R_{min} = 0.414$ in Störmer units C_{st}.

A number of equatorial orbits for various values of γ are illustrated in Figs. 2 and 3.

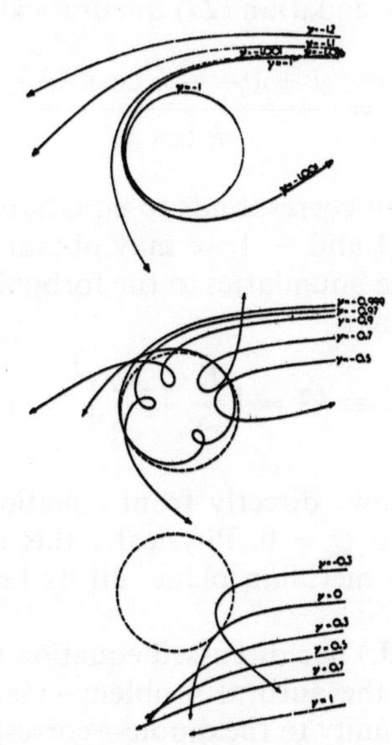

Fig. 2. Störmer trajectories in the equatorial plane. The circle has a radius of one Störmer unit, C_{st}. After Störmer (1955); courtesy Oxford University Press.

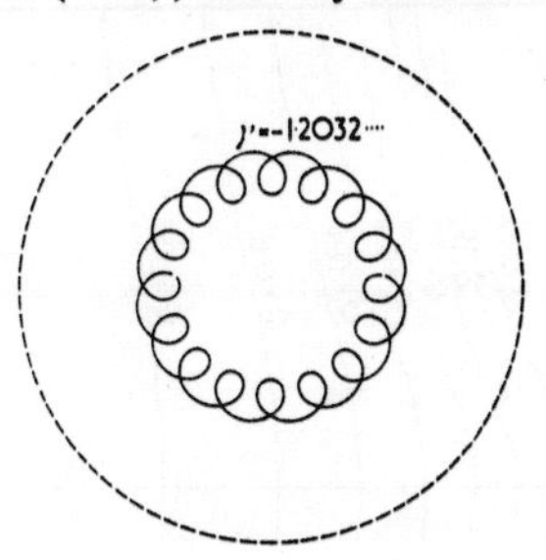

Fig. 3. An example of a captive orbit in the equatorial plane. For this particular value of γ, the orbit is periodic, in that it repeats itself precisely with each revolution about the dipole axis. After Störmer (1955); courtesy Oxford University Press.

1.5 Forbidden Regions in the Three-Dimensional Problem

To generalize the treatment in the preceding section, let ω be the angle between the tangent to the orbit and the meridian plane, and let λ be the latitude of the particle at a given instant. In the meridian plane the particle's velocity is then $\cos \omega$ and transverse to this plane it is $R d\varphi/ds = \sin \omega$.

Refer now to the angular-momentum equation (13), which is the basis for establishing the forbidden regions. Since $R = r \cos \lambda$, we have

$$r \cos \lambda \cdot \sin \omega = \frac{\cos^2 \lambda}{r} + 2\gamma. \tag{27}$$

If we set $k \equiv \sin \omega$ and solve equation (27) quadratically, we obtain

$$r = \frac{\gamma \pm (\gamma^2 + k \cos^3 \lambda)^{1/2}}{k \cos \lambda}, \tag{28}$$

where $|k| \leqslant 1$. This equation corresponds to equation (26) in the two dimensional problem. By setting $k = +1$ and -1 we may obtain the points at which the particles turn around, that is, the boundaries to the forbidden regions. It will be noticed from equation (13) or (27) that

$$\sin^2 \omega \equiv k^2 = \left(\frac{R}{r^3} + \frac{2\gamma}{R}\right)^2 = 1 - Q, \tag{29}$$

where the last equality follows directly from equation (14). Hence $Q = \cos^2 \omega$, and $k = \pm 1$ corresponds to $Q = 0$. Physically this means simply that when the particle turns around in the meridian plane, all its kinetic energy is in transverse motion.

Störmer (1955, pp.232 ff.) has discussed equation (28) in detail and has shown that the case of interest in the auroral problem—vix., when the allowed regions stretch continuously from infinity to the dipole—corresponds to values of γ between 0 and -1.

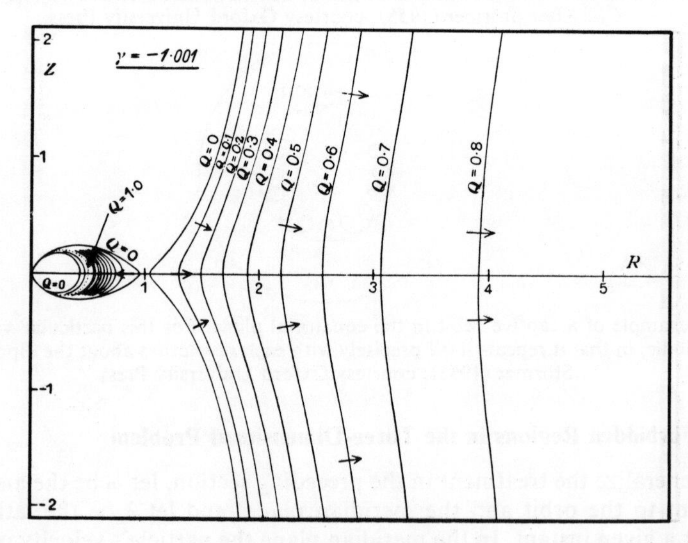

Fig. 4. Regions of constant Q in the meridian plane for $\gamma = -1.001$. After Störmer (1955); courtesy Oxford University Press.

Figs. 4 and 5 illustrate the regions of constant Q within the meridian plane for values of γ on either side of $\gamma = -1$. These Q curves are obtained from equation (28), where k is related to Q by equation (29). For our present discussion, the important point to notice in the figures is that a $Q = 0$ curve forms a barrier for a particle. Thus a particle coming from the sun (at large R), with $\gamma = -1.001$. cannot penetrate to the dipole (i.e., to the origin). On the other hand, if $\gamma = -0.999$, the particle can just squeeze through the neck around $R = 1$. It can reach the dipole, if it is projected just right, by sliding toward the polar regions between the two $Q = 0$ curves.

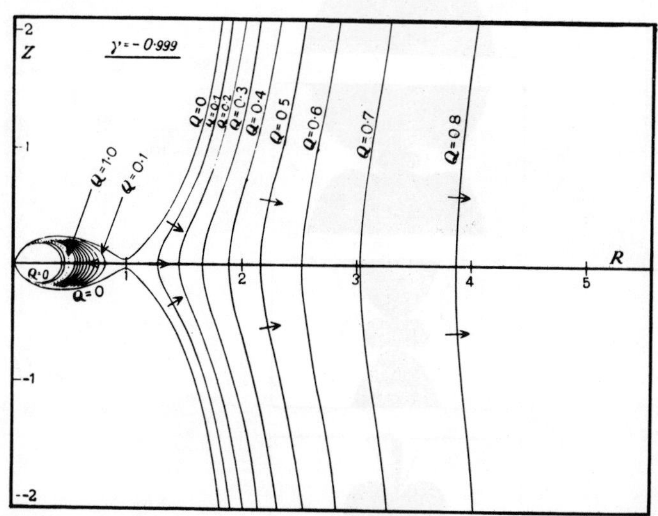

Fig. 5. Regions of constant Q in the meridian plane for $\gamma = -0.999$. After Störmer (1955); courtesy Oxford University Press.

For protons with sufficient energy to penetrate to auroral depths (5×10^5 ev), the Störmer unit of length is 3×10^{10} cm, nearly the radius of the moon's orbit. Hence the earth, with a radius of 6.4×10^8 cm, would be quite small in Figs. 4 and 5. This explains our earlier statement, that it is necessary to consider only those orbits that can actually penetrate to the origin.

As γ increases above -0.999 the neck at $R = 1$ opens wider and makes it easier for particles to slip in. But at the same time the inner $Q = 0$ region increases in size, so that when $\gamma > 0$ the dipole is again completely blocked, as illustrated in Fig. 6.

1.6 Motion of a Particle in Three Dimensions

We consider separately the motion *within* the meridian plane and the (nonuniform) rotation of this plane about the z axis. In some respects this method of dividing the problem can be a little confusing, when we try to visualize the orbit as a whole, but it is the most convenient way to attack the mathematical problem.

First consider the motion of the meridian plane. We have shown that for auroral practical purposes $\gamma < 0$. Hence define $\gamma_1 \equiv -\gamma$, where γ_1 will always be a positive number. The angular-momentum integral (13) is thus

$$\frac{d\varphi}{ds} = -\frac{2\gamma_1}{r^3 \cos^2 \lambda}\left(r - \frac{\cos^2 \lambda}{2\gamma_1}\right), \tag{30}$$

since $R = r \cos \lambda$. Hence $d\varphi/ds$ changes sign and the meridian plane reverses its

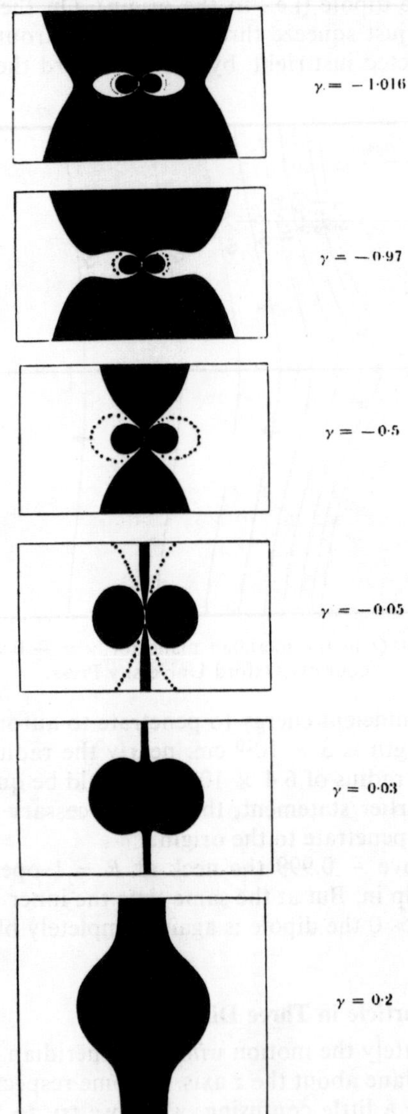

Fig. 6. The diagrams show (in white) allowed regions for different values of γ. The figure illustrates why only values of γ between -1 and 0 are important to the auroral trajectories. After Störmer (1955); courtesy Oxford University Press.

direction of motion when

$$r = \frac{\cos^2 \lambda}{2\gamma_1}. \tag{31}$$

It may easily be shown from the r and λ components of a dipole field, that the equation of a line of force has the same form as equation (31). Thus a particle reverses direction when it reaches the surface formed by rotating the line of force (31) about the z axis.

Referring to Figs. 2 and 3 we see that this situation arises in the equatorial plane when the orbit forms a little loop. On the inside portion of the loop the φ component of motion is in the opposite direction to that over the main portion of the curve.

In the three-dimensional case, as a particle spirals toward the dipole around a line of force (approximately), its meridian plane reverses twice during each loop of the spiral. This phenomenon occurs because near the dipole the guiding center of the particle follows closely a line of force. Picture the surface formed by rotating this line of force about the z axis. Then every time the particle crosses this surface the meridian plane reverses. Farther from the dipole, the guiding center diverges from the line of force, but the reversal still occurs precisely when the particle crosses this surface.

It is also of interest that whenever the meridian plane reverses direction ($d\varphi/ds = 0$), we find, by comparing equations (13) and (14) that $Q = 1$. From the second equality of equation (14), it is apparent that this Q is consistent with all the kinetic energy being in the meridian plane and none in the φ direction. Hence, by the discussion in the previous paragraph, we see that a $Q = 1$ curve corresponds to a line of force in the (moving) meridian plane.

The motion within the meridian plane is governed by equations (17) and (18). These equations are analogous to the equations of motion in the R, z plane of a free particle of unit mass that rolls over hills and valleys in the third dimension. In this analogy we consider s as representing time and $-(\frac{1}{2})\,Q$ as the potential energy.

From Fig. 5 it becomes clear that as a particle coming from infinity approaches the neck, and again as the particle approaches the dipole, the curves for constant Q crowd closer and closer together. In general, in the neighborhood of small Q, the particle experiences a force tending to deflect it back. Where the Q values crowd together this repulsive force increases. The total speed remains constant, so energy lost from motion in the meridian plane goes into angular motion of the plane. As the particle moves back to larger Q values the φ motion and the repulsive force decrease.

Now consider a particle moving toward the dipole and spiraling about a line of force. It will continually interchange energy between its different components of motion. At two points of each loop on the spiral it will have all its motion in the meridian plane ($Q = 1$) and at two other points (about 90° from the first two) a large portion of the motion is in the transverse direction (small Q). Unless the particle is projected just right to reach the dipole, it will eventually, at some point on its orbit, have all its energy in the transverse direction ($Q = 0$). When this occurs the particle starts to spiral out again. Whereas this discussion for the case of the dipole is necessarily a little complicated, it is merely the exact dipole solution of the problem of magnetic reflection, discussed more generally in Section 2.

The actual computation of three-dimensional orbits has been carried out numerically by Störmer and others. Our discussion here has merely attempted to show some of the characteristics of the orbits and clarify the physical meaning of the differential equations involved.

1.7 Orbits Through the Origin; Families of Orbits

We have seen that the characteristics of an orbit may be studied from the standpoint of the Q value of the particle at different points on its orbit. Thus the motion of the meridian plane reverses whenever $Q = 1$, and the particle is magnetically reflected when $Q = 0$. The curve $Q = 1$ is merely a particular line of force (depending on the value of $\gamma_1 \equiv -\gamma$) in the meridian plane. On the other hand, the $Q = 0$ curves may be obtained from equation (14). Substituting $R = r \cos \lambda$ and solving the resulting equation, quadratic in $1/r$, we find for $Q = 0$ that

$$r = \frac{\cos^2 \lambda}{\gamma_1 + (\gamma_1^2 \pm \cos^3 \lambda)^{1/2}}. \tag{32}$$

In obtaining this solution, we chose only the positive sign in front of the square root, since asymptotically near the dipole these curves must merge into the $Q = 1$ curve (or line of force) given by equation (31). The plus-minus sign in equation (32) arises from the squared term in equation (14) and gives two $Q = 0$ curves.

Fig. 7 shows the curves for $Q = 0$ and 1 near the origin for $\gamma_1 = 0.5$. Notice that the numerical computations of Störmer show that the orbit through the origin is not precisely along a line of force. This is understandable when we recall that a particle moving in an inhomogeneous field will drift (in the φ direction). Hence, to project a particle so that it will go into the origin asymptotically with a line of force, we should start it with just enough angular motion to compensate for the drift. This means that initially $Q \neq 1$; but this orbit through the origin approaches $Q = 1$ as the dipole is approached.

Let us now visualize a group of orbits all starting from (or passing through) a given point R_0, z_0, φ_0. This point will lie on an orbit through the origin for some particular value of γ_1. Hence, let us consider only those particles coming from this point that have this value of γ_1 (or initial angular momentum). (Having determined both meridional coordinates of the point, we cannot independently specify both γ_1 and the condition that the point is somewhere on the orbit through the origin.)

The orbit is completely fixed by five parameters. (The general equation of motion is a sixth order equation, the final integration constant fixing the particle in the orbit at a particular time.) So far we have specified γ_1, and fixed z and φ as functions of R_0. As a fourth initial condition we specify the total velocity or the unit of length C_{st}; these four conditions specify the "family". As we vary the fifth initial condition (which specifies the velocity component in some direction in the meridian plane) we obtain the various members of the family, three of which are shown in Fig. 7.

These various orbits oscillate within the meridian plane until they touch a $Q = 0$ barrier, then recede back to infinity. One should also visualize the meridian plane as oscillating and drifting in the φ direction, with maximal angular velocities for $Q = 0$ and zero velocity when the particle is at $Q = 1$.

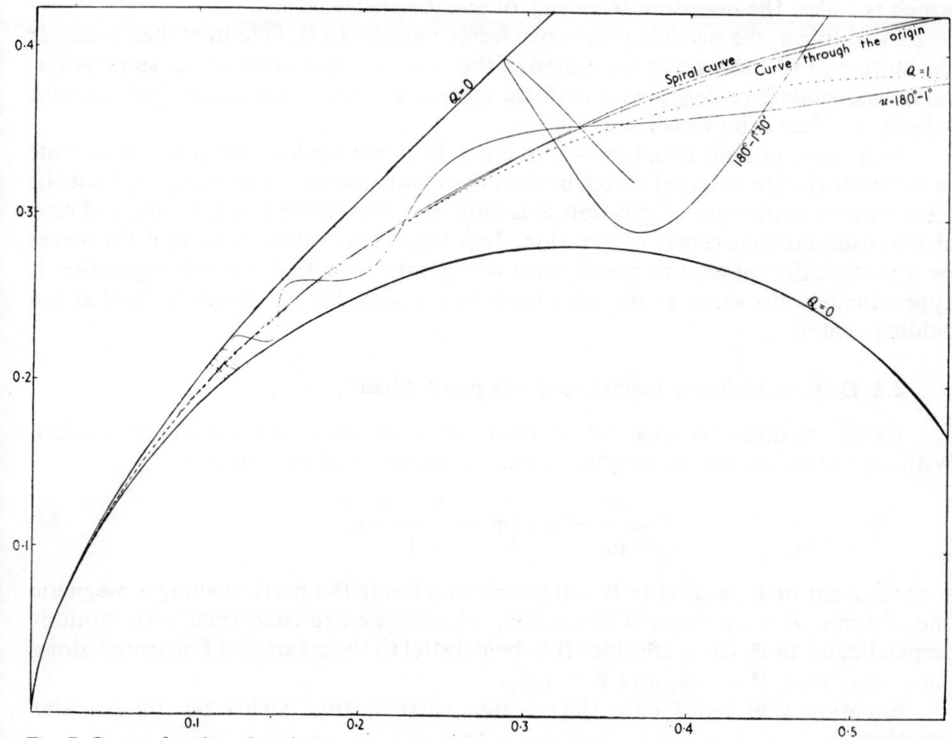

Fig. 7. Curves for $Q = 0$ and $Q = 1$ near the dipole. The diagram shows a section of the meridian plane: Abscissa and ordinate are the R and z axes, respectively. The orbit through the origin and various spiral orbits are illustrated. After Störmer (1955); courtesy Oxford University Press.

A family member (labelled "spiral curve" in Fig. 7) just off the orbit through the origin is of interest. It oscillates about the curve through the origin, *not* about the line of force. Hence its φ motion varies in magnitude but the meridian plane does not reverse. For such a particle the drift motion outweighs the spiralling, and the former merely appears slightly irregular.

2. Guiding-Center Theory of Particle Orbits

A charged particle in a uniform magnetic field, obeying the equation of motion (1), being accelerated perpendicular to **v** and to **B** circles a magnetic line of force. This gyration is in the direction such that the small field *produced by* the particle is in the direction opposite to the external field.

Let $v_\perp$ be the velocity component perpendicular to **B**, and let ρ be the radius of gyration. Equating the centrifugal force, $mv_\perp^2/\rho$, to the Lorentz force, $(e/c)\, v_\perp B$, we find an angular velocity,

$$\omega_0 \equiv \frac{v_\perp}{\rho} = \frac{eB}{mc}, \tag{33}$$

which is called the *cyclotron frequency* or *gyrofrequency*.

In addition, the particle may move freely parallel to **B**. One may then consider the motion of the particle as composed of the sum of its gyrations about some point, called the *guiding center*, plus a uniform motion of this center along the field with velocity $v_{||}$. The orbit thus traces a helix.

In a more general situation where the field is not uniform or constant in time or where there are external forces besides the magnetic field, one may still treat the total motion as the sum of gyration plus motion of the guiding center—defined now as the instantaneous center of gyration. This type of treatment is valid if the forces on a particle do not change greatly over one gyration, so that the actual gyration is approximately the same as the uniform-field gyration for the magnetic field at the guiding center.

2.1 Drift in Uniform Electric and Magnetic Fields

As an introductory case, let us treat the situation of crossed uniform fields. With an electric as well as magnetic field the equation of motion is

$$m\frac{d\mathbf{v}}{dt} = \pm e\left(\mathbf{E} + \frac{\mathbf{v} \times \mathbf{B}}{c}\right). \tag{34}$$

A component of **E** parallel to **B** will simply accelerate the particle along a magnetic line of force. For the present discussion, wherein we are concerned with motions perpendicular to **B**, let us consider **B** to be parallel to the z axis and **E** oriented along the y axis. Thus $\mathbf{B} = \mathbf{i}_z B_z$ and $\mathbf{E} = \mathbf{i}_y E_y$.

Equation (34) (with only the + sign carried) thus yields the three scalar equations

$$\dot{v}_x = v_y \omega_0, \tag{35}$$

$$\dot{v}_y = \frac{eE_y}{m} - v_x \omega_0, \tag{36}$$

and

$$\dot{v}_z = 0, \tag{37}$$

where we now use dots to denote time derivatives.

We assume a solution of a form that will describe both a gyrational motion and any additional drifting:

$$v_x = v_\perp \sin \omega_0 t + V_x, \tag{38}$$

where $v_\perp$ is the instantaneous amplitude of gyration and V_x is an additional velocity. From equation (35) we have

$$v_0 = v_\perp \cos \omega_0 t + \frac{\dot{v}_\perp}{\omega_0} \sin \omega_0 t + \frac{\dot{V}_x}{\omega_0}. \tag{39}$$

Then neglecting second-order derivatives of the velocity, we have from equation (36)

$$2\dot{v}_\perp \cos \omega_o t = \frac{eE_y}{m} - V_x\omega_o. \tag{40}$$

Since this is an identity, valid for any t, we have $\dot{v}_\perp = 0$ and

$$V_x = \frac{eE_y}{m\omega_o} = \frac{cE_y}{B_z}. \tag{41}$$

This is the "$\mathbf{E} \times \mathbf{B}$ drift". The description of the motion as the sum of gyration plus a uniform drift, as given by equations (38) and (39) (with $\dot{v}_\perp = \dot{V}_x = 0$) is exact. By writing the total velocity as $\mathbf{v} = \mathbf{v}' + \mathbf{V}$, where

$$\mathbf{V} = \frac{c\mathbf{E} \times \mathbf{B}}{B^2}, \tag{42}$$

one may readily show from equation (34) that $\mathbf{v}'$ represents only gyrations (in a frame of reference moving with velocity $\mathbf{V}$). From a different point of view, we may note, following Alfvén, that an electric field $\mathbf{E}$ in the "laboratory" frame of reference will be relativistically transformed to zero in a frame of reference moving with the velocity $\mathbf{V}$ given by equation (42). In that frame, the particle sees no electric field and merely performs gyrations about the magnetic field.

2.2 General Theory of Particle Drifts

Electric forces are not the sole cause of particle drifts. Any similar force will produce drifting. However, it is of interest that all particles, regardless of mass or electric charge, drift in the same direction and with the same speed under the influence of crossed electric and magnetic fields; this may be regarded as a simple consequence of the relativistic Lorentz transformation. In the case of a gravitational field, the positively and negatively charged particles drift in opposite directions. Similarly a particle moving along curved lines of force experiences a centrifugal acceleration, which is equivalent to a gravitational field in producing a drift across lines of force. In this section we shall thus treat an external force that does not depend on the electric charge.

A rather different type of perturbation is that caused by the non-uniformity of the magnetic field. For an introductory treatment, we shall take the magnetic gradient parallel to an external force and both these perpendicular to an electric field (see Fig. 8). This is sufficiently general to show the effect of a drift motion that is parallel to another force. (A more general treatment is given by Northrup 1961; also see Kruskal 1960b. In general, it is necessary to assume that the longitudinal electric field is very weak, so that the rate of energy gain over a gyration is small.)

Hence we take $\mathbf{B} = \mathbf{i}_z B_z$ with B_z dependent on position: $B_z = B_0 + B'(x - x_0)$, where $B' = \partial B_z/\partial x$. The electric field is $\mathbf{E} = \mathbf{i}_y E_y$ and another external force is $m\mathbf{f} = m\mathbf{i}_x \mathbf{f}_x$. We shall now write $\omega_0 = eB_0/mc$ and $\omega' = eB'/mc$.

With an external force that is independent of charge the equation of motion is

$$\frac{d\mathbf{v}}{dt} = \pm \frac{e}{m}\left(\mathbf{E} + \frac{\mathbf{v} \times \mathbf{B}}{c}\right) + \mathbf{f}. \tag{43}$$

We shall again carry only the + sign in this equation, applicable to positively charged particles. At any stage in the development the insertion of a minus sign wherever the charge e appears will alter the equations to negative charges.

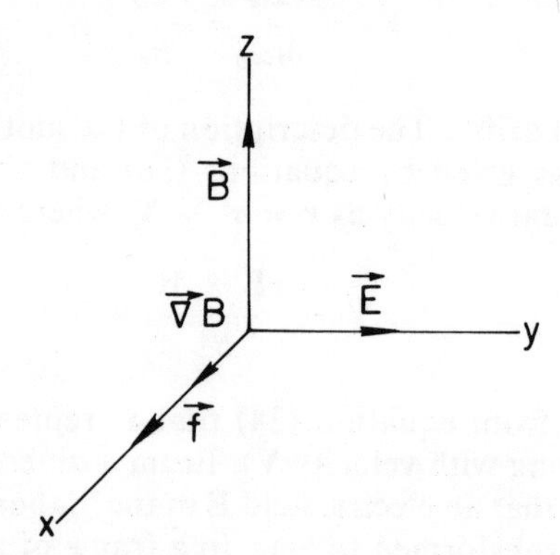

Fig. 8. Orientations of vectors as treated in the text.

For the particular orientation of forces described above, equation (43) yields

$$\dot{v}_x = v_y[\omega_0 + \omega'(x - x_0)] + f_x \tag{44}$$

and

$$\dot{v}_y = \frac{eE_y}{m} - v_x[\omega_0 + \omega'(x - x_0)]. \tag{45}$$

We assume a first-order solution composed of a circular motion with the gyro-frequency and superimposed drift velocities:

$$v_x = v_\perp \sin \omega_0 t + V_x \tag{46}$$

and

$$v_y = v_\perp \cos \omega_0 t + V_y. \tag{47}$$

Integration of equation (46) for a constant amplitude, $v_\perp$, gives

$$x - x_0 = -\frac{v_\perp}{\omega_0}\cos \omega_0 t + V_x t. \tag{48}$$

Similarly equations (46) and (47) may be differentiated. When the various terms have been substituted in equations (44) and (45) we obtain

$$v_\perp \omega_0 \cos \omega_0 t = (v_\perp \cos \omega_0 t + V_y)\left[\omega_0 + \omega'\left(-\frac{v_\perp}{\omega_0}\cos \omega_0 t + V_x t\right)\right] + f_x \tag{49}$$

and

$$-v_\perp\omega_0 \sin \omega_0 t = \frac{e}{m}E_y - (v_\perp \sin \omega_0 t + V_x)\left[\omega_0 + \omega\left(-\frac{v_\perp}{\omega_0}\cos \omega_0 t + V_x t\right)\right]. \quad (50)$$

The cyclic terms of amplitude $v_\perp\omega_0$ on either side of these two equations cancel. The remaining terms govern the drifts. Averaging the equations over a cycle around $t = 0$, the only periodic term which contributes to a net drift is that containing $\cos^2 \omega_0 t$ in the expanded equation (49). Thus

$$V_y = \frac{\omega' v_\perp{}^2}{2\omega_0{}^2} - \frac{f_x}{\omega_0} \quad (51)$$

and

$$V_x = \frac{eE_y}{m\omega_0}. \quad (52)$$

Equation (52) is identical to (41) and merely represents the same $\mathbf{E} \times \mathbf{B}$ drift we obtained when there were no other external forces. Equation (51) gives a combined drift from the gradient of the field perpendicular to itself and from the external force. In general, the ∇B drift is in the direction $\pm\, \mathbf{B} \times \nabla B$ and that due a force per unit mass, $\mathbf{f}$, is toward $\pm\, \mathbf{f} \times \mathbf{B}$.

Let us now consider the particular drift of a particle moving along curved lines of force. The particle experiences a virtual (centrifugal) force by being constrained to a noninertial coordinate system. Then $\mathbf{f} = v_{||}{}^2/\mathcal{R}$, where $\mathcal{R}$ is the radius of curvature of the field lines. In a region in which there are no currents flowing, Ampère's law requires that $\nabla \times \mathbf{B} = 0$. This condition may be written as

$$\frac{dB}{dR} = -\frac{B}{\mathcal{R}}, \quad (53)$$

where the coordinate R is directed outward along the radius of curvature. Hence we write $f = -v_{||}{}^2\omega'/\omega_0$ and the drift due to the field gradient and that due to the curvature of the lines of force are in the same direction. The total drift is parallel to $\pm\, \mathbf{B} \times \nabla B$ or $\pm\, \mathbf{R} \times \mathbf{B}$ and has a magnitude

$$V_y = \frac{\omega'}{\omega_0{}^2}(\tfrac{1}{2}v_\perp{}^2 + v_{||}{}^2) = \frac{\nabla_\perp B}{\omega_0 B}(\tfrac{1}{2}v_\perp{}^2 + v_{||}{}^2). \quad (54)$$

In the earth's field a positive charge drifts westward, a negative one eastward, thereby producing a westward current system.

2.3 The Transverse Adiabatic Invariant: The Magnetic Moment[3]

Thus far in considering magnetic fields we have restricted ourselves to lines of force that are always parallel to one another, whether they be curved or straight. However a most interesting effect appears when we follow a charged particle in a field where the lines of force converge toward one another.

[3]Based in part on a review by Chamberlain (1961). Section 3.2.4.

Picture a particle with a spiral path symmetric about a line of force on the z-axis. This trajectory is composed of a gyrational velocity v_φ and a motion of the *guiding center* along the field with velocity v_z. Off the z-axis the magnetic field has a small component B_R measured positive away from the z-axis. The field is azimuth-independent and $B_\varphi = 0$.

Since the lines of force must be continuous, $\nabla \cdot \mathbf{B} = 0$ or

$$\frac{1}{R}\frac{\partial}{\partial R}(RB_R) + \frac{\partial B_z}{\partial z} = 0. \tag{55}$$

The convergence is assumed to be gradual, so that during the time required for the particle to make a single gyration it has experienced little change in the field. Then we may set $\partial B_z/\partial z \approx \partial B/\partial z$. Integrating equation (55), we have

$$RB_R = -\frac{R^2}{2}\frac{\partial B}{\partial z}. \tag{56}$$

At the position of the particle ($R = \rho$), we have

$$B_R = -\frac{\rho}{2}\frac{\partial B}{\partial z}. \tag{57}$$

If we consider the possibility of a uniform electric field E_z accelerating the particle along the magnetic field, the equation of motion (34) is

$$m\frac{dv_z}{dt} = \pm\, e\left(E_z + \frac{1}{2}\frac{\rho v\varphi}{c}\frac{\partial B}{\partial z}\right). \tag{58}$$

Employing equation (33) to eliminate ρ, we have

$$m\frac{dv_z}{dt} = \pm\, eE_z - \frac{m}{2}\frac{v_\phi{}^2}{B}\frac{\partial B}{\partial z}, \tag{59}$$

where the minus sign is inserted in the second term on the right because the sense of gyration of a positive particle is in the $-\varphi$ direction with our convention for the direction of the field (i.e., $v_\varphi = \mp v_\perp$). The equation is valid for particles of either sign.

If the lines of force are converging in the direction of motion ($\partial B/dz > 0$), the magnetic field tends to decelerate the forward motion of the particle. However as we have shown in equation (2), the magnetic field alone cannot change the total speed of a particle. Hence, it is clear that the velocity lost along the field must reappear as an increase in the absolute value of the v_ϕ component. We shall investigate this point further.

It is convenient to write the equation of motion (59) in terms of the magnetic moment of the particle, μ, which is defined as the product of the current produced by the particle times the area encircled by the current. Thus

$$\mu \equiv \frac{e\omega_0}{2\pi c}\cdot \pi\rho^2 = \frac{mv_\perp{}^2}{2B}, \tag{60}$$

where the second equality follows directly from equation (33).

Multiplying the equation of motion (59) by v_z and substituting μ from equation (60), we obtain

$$\frac{d}{dt}\left(\frac{1}{2}mv_z^2\right) = \pm eE_z v_z - \mu\frac{dB}{dt}. \tag{61}$$

Here d/dt indicates the substantial derivative, which is taken along the path of the particle. (For a stationary observer, we would have $\partial B/\partial t = 0$.)

Another relation between v_z and μ (or v_φ) can be found from energy considerations. Since the total kinetic and potential energy of the system is a constant,

$$\frac{d}{dt}\left(\frac{1}{2}mv_z^2 + \frac{1}{2}mv_\phi^2\right) = \pm eE_z\frac{dz}{dt} = \pm eE_z v_z. \tag{62}$$

Using equation (6) we then obtain

$$\frac{d}{dt}\left(\frac{1}{2}mv_z^2\right) = \pm eE_z v_z - \frac{d}{dt}(\mu B). \tag{63}$$

A comparison of equation (61) and (63) illustrates that, in the limit of our approximation of a slowly converging field, μ is a constant.

Let θ be the *pitch angle* between the total velocity vector $\mathbf{v}$ and the magnetic field. Then $v_\varphi = v \sin \theta$. Further, let

$$\epsilon = \frac{1}{2}m(v_z^2 + v_\phi^2) = \frac{1}{2}mv^2,$$

the kinetic energy of the particle.

The constancy of the magnetic moment may then be expressed as

$$\mu \equiv \frac{mv_\phi^2}{2B} = \frac{mv^2 \sin^2\theta}{2B} = \frac{\epsilon \sin^2\theta}{B} = \text{const.} \tag{64}$$

If there is no electric field acting on the particle, so that ϵ = constant, and at a given point on the trajectory the field and angle of pitch are B_1 and θ_1, respectively, then the particle will be *magnetically reflected* when the field seen by the particle increases to the value

$$B_m = \frac{B_1}{\sin^2\theta_1}, \tag{65}$$

where B_m defines a *magnetic mirror* point.

At this point all the kinetic energy has been transformed into the gyration of the particle. But it is clear from equation (59) that so long as $\partial B/\partial z > 0$, there will be a force on the particle in the $-z$ direction so that the particle recedes, gaining speed parallel to the field as $|v_\varphi|$ decreases.

If there is an electric field involved, the last equation in the set (64) should be used, rather than the more familiar relation (65). Again we must caution that these relations are not exact and do not apply strictly, for example, to the motion of particles over large distances in the field of a dipole. Nevertheless, Alfvén (1950) has successfully applied equation (64), along with equation (54) for the perpendicular drift of the guiding center, to an approximate treatment of the Störmer problem—the motion of a charged particle in the field of a dipole.

Equation (64) is sometimes called the *first* or *transverse adiabatic invariant* and was first developed and applied in the general case by Alfvén (1950) (also see Landau and Lifshitz 1951).

The magnetic moment is also conserved during slow changes of the field B with time (Spitzer 1956) and for drift motions that carry a particle into a different magnetic field. To prove the latter statement, we substitute the first-order solutions (46) and (47) in the equations of motion (44) and (45) and average over a cycle:

$$\frac{m}{2}\frac{dv_\perp^2}{dt} = \langle mv_x\dot{v}_x + mv_y\dot{v}_y\rangle = eE_yV_y + mV_xf_x. \tag{66}$$

Substituting for V_y from equation (51) and then for E_y from equation (52) we have

$$\frac{m}{2}\frac{dv_\perp^2}{dt} = \frac{mv_\perp^2}{2\omega_0}\omega' V_x = \frac{mv_\perp^2}{2\omega_0}\frac{d\omega_0}{dt}, \tag{67}$$

where $d\omega_0/dt$ is the substantial derivative measured along the trajectory of the guiding center. From the definition (60) of μ and with relation (33), equation (67) may be written

$$\frac{d\log\mu}{dt} = 0. \tag{68}$$

A particle that is caused (by a non-uniform magnetic field) to drift parallel to an electric field will clearly gain or lose kinetic energy, as its electric potential decreases or increases. However, the above deriviation shows that there will be a simultaneous drift (owing to the electric field) toward a stronger or weaker magnetic field, which is exactly sufficient to maintain a constant magnetic moment for the particle.

The constancy of μ is an enormous aid in treating the motions of trapped particles, since it relates the mirror points to the pitch angles anywhere else in the field.

The conservation of the magnetic moment may be regarded as a consequence of the conservation of a particle's angular momentum about the magnetic field. The latter is

$$mv_\perp\rho = m^2v_\perp^2c/eB = \frac{2mc}{e}\mu, \tag{69}$$

where we have used equation (33) to eliminate ρ.

Since the magnetic moment μ is aligned anti-parallel to the external field, a plasma is often said to be diamagnetic, although the usual magnetic susceptibility is not a constant. The magnetization, or magnetic moment per unit volume is

$$\mathbf{M} = -\frac{\mathscr{E}_\perp}{B^2}\mathbf{B}, \tag{70}$$

where $\mathscr{E}_\perp$ is the transverse kinetic energy per unit volume. The magnetic induction within a plasma is then $\mathbf{B} = \mathbf{B}_0 + 4\pi\mathbf{M}$, where $\mathbf{B}_0$ is the magnetic field in the absence of a plasma. Thus the magnetic energy becomes, for $B \gg M$,

$$\frac{B^2}{8\pi} = \frac{B_0{}^2}{8\pi} - \mathscr{E}_\perp, \tag{71}$$

showing that the sum of the magnetic and transverse kinetic energies (and therefore the sum of their transverse pressures as well) remains constant as the kinetic energy is added.

A magnetic field may thus confine a plasma imbedded within it, although such a system happens to be subject to instabilities that lie beyond the scope of this article.

2.4 The Longitudinal Adiabatic Invariant

A second adiabatic invariant of a particle's motion is the longitudinal or integral invariant. Its principal usefulness in geomagnetic problems lies in predicting the surface swept out by drifting particles. If there were no electric fields and the magnetic field arose from a perfect dipole, this problem would be trivial—the surfaces would be those created by rotating lines of force about the axis.

For an unsymmetrical magnetic field one may see qualitatively the deviation from such a simple surface by considering a particle confined to the equatorial plane—that is, one with a large pitch angle. The magnetic drift is parallel to $\mathbf{B} \times \nabla B$ and the particle thus drifts on a line of constant B and returns to the same point after one revolution around the earth. If the particle starts from a region where the dipolar field is compressed (that is, increased in strength) by say, the pressure of a solar wind it will drift closer to the earth, as it enters an undisturbed region of the field, so as to remain on a constant B.

For the general case a particle moves in such a way that the action integral,

$$\mathrm{J} = \oint mv_{\parallel} ds, \tag{72}$$

is a constant of the motion, where ds is measured along the field lines and the integral is evaluated over a line of force between two mirror points of the orbit (see Fig. 9).

In the special case where v = constant, equation (65) is valid and

$$I = \oint \left(1 - \frac{B}{B_m}\right)^{1/2} ds \tag{73}$$

is constant. Vestine and Sibley (1959) have made sample calculations of I and applied them to predictions of the shape of the auroral zone.

Integrating the energy equation (63) for static electric and magnetic fields, we have

$$\frac{1}{2} mv_{\parallel}{}^2 + \mu B + e\Phi = H, \tag{74}$$

where Φ is the electrostatic potential and H is the total energy, which is constant. (When H is constant, μ, J, and H constitute the three principal constants of motion.) Then equation (72) may be written as

$$J = \oint [2m(H - e\Phi - \mu B)]^{1/2} ds. \tag{75}$$

Being an integral over a line of force, J is not altered by variations parallel to **B**, but could conceivably be affected by transverse drifts.

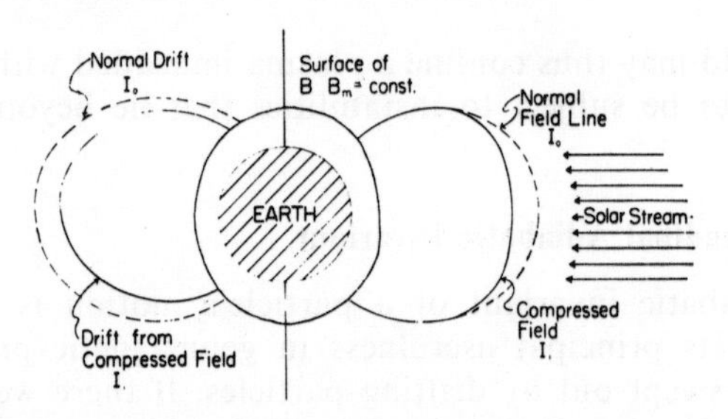

Fig. 9. Drift of trapped particles according to the longitudinal invariant. Solar gas incident from the right compresses the field and injects particles with mirror points on the surface of constant B. In drifting to the night side, the particles must move even closer to the earth to conserve I.

Here we shall prove the constancy of J for static fields, in a geometrical fashion that will serve to illustrate the physics involved. Northrup and Teller (1960) have given a relativistic general proof, including slow time variations in **E** and **B**. Their description of the spatial variations is rather different from—but nevertheless entirely equivalent to—that given here.

Let us suppose that the guiding center of a particle drifts from one field line to another adjacent line while the particle is at a specified segment, ds, of the field line. The change in the integral (75) is affected not only by the changes occurring at the position of the particle but also must include the variations at every other element ds' along the entire line of force. When the particle is at the segment ds the motion of the line segment ds' need not necessarily be in the same direction as when the particle itself is at ds'. The principal difficulty in the proof will lie in evaluating the motion at ds' when the particle is at ds.

Differentiating equation (75) gives

$$\frac{dJ}{dt} = -\oint \left(\frac{m}{2}\right)^{1/2} \frac{(ed\Phi/dt + \mu dB/dt)'}{(H - e\Phi - \mu B)'^{1/2}} ds' + \oint [2m(H - e\Phi - \mu B)']^{1/2} d\left(\frac{ds'}{dt}\right), \tag{76}$$

where a prime on any quantity means it is evaluated at s'. The second integral expresses the change in arc length as a particle drifts from one line to another of different length. The elements ds' on the two lines are in direct proportion to the radii of curvature of these lines. We find from simple geometrical considerations

$$\delta(ds') = \delta\mathscr{R}\frac{ds'}{\mathscr{R}} \tag{77}$$

or, since $\nabla_{\perp}B/B = -1/\mathscr{R}$ by equation (53),

$$\mathrm{d}\left(\frac{ds'}{dt}\right) = -\mathbf{W}'\cdot\left(\frac{\nabla B}{B}\right)' ds, \tag{78}$$

where $\mathbf{W}'$ represents the virtual rate of displacement (perpendicular to (**B** of the line of force at s', owing to the particle drifting at s.

Equation (76) may now be written as

$$\frac{dJ}{dt} = -\oint\frac{ds}{v_{\parallel}'}\mathbf{W}'\cdot\left(e\nabla\Phi + \mu\nabla B + \frac{mv_{\parallel}^2}{B}\nabla B\right)'. \tag{79}$$

With $\mathbf{E} = -\nabla\Phi$, it is clear from equations (42) and (54) that the terms in parenthesis all appear in the drift velocity **V**. The latter is

$$\mathbf{V} = \frac{c}{e}\frac{\mathbf{B}}{B^2}\times\left[-e\mathbf{E} + \nabla B\left(\frac{mv_{\perp}^2}{2B} + \frac{mv_{\parallel}^2}{B}\right)\right]. \tag{80}$$

Taking the vector product of **B** with equation (80) we find, for equation (79),

$$\begin{aligned}\frac{dJ}{dt} &= -\frac{e}{c}\oint\frac{ds'}{v_{\parallel}'}\mathbf{W}'\cdot(\mathbf{V}'\times\mathbf{B}') \\ &= \frac{e}{c}\oint\frac{ds'}{v_{\parallel}'}\mathbf{B}'\cdot(\mathbf{V}'\times\mathbf{W}').\end{aligned} \tag{81}$$

We must now find the product of $\mathbf{V}'$, the drift velocity of a particle at s', with $\mathbf{W}'$, the rate at which the line attached to a particle at s is shifting at the point s'. Let us represent a line of force by parametric equations in which position along the line is the independent variable: $\mathbf{r} = \mathbf{r}(s; \alpha_i)$. Here α_i (for $i = 1, 2 \ldots n$) represents the various n parameters required to specify a particular line of force. For example, a dipole line of force may be specified by $\varphi = a$ and $r = b\cos^2\lambda$, where φ is longitude and λ is latitude. Then a and b would be the parameters α_1 and α_2. A displacement of position of a point **r** is then

$$d\mathbf{r} = \frac{\partial\mathbf{r}}{\partial s}ds + \sum_{i=1}^{n}\frac{\partial\mathbf{r}}{\partial\alpha_i}d\alpha_i. \tag{82}$$

The first term on the right gives a displacement along the line of force, and in the following we shall ignore it, being concerned here with shifts perpendicular to the line. Thus, when the particle is located at the point s', we have

$$\frac{d\mathbf{r}}{dt} = \mathbf{V}' = \sum_i\left(\frac{\partial\mathbf{r}}{\partial\alpha_i}\right)'\left(\frac{d\alpha i}{dt}\right). \tag{83}$$

When the particle is at s, the point at s' is displaced with a rate

$$\frac{d\mathbf{r}'}{dt} = \mathbf{W}' = \sum_j \left(\frac{\partial \mathbf{r}}{\partial \alpha_j}\right)' \frac{d\alpha_j}{dt}, \tag{84}$$

where the total derivative on the right is evaluated from the rate of particle drift at s.

In the matrix of terms formed by the product $\mathbf{V}' \times \mathbf{W}'$ the diagonal $(i = j)$ terms clearly vanish and the others occur in pairs such that the expanded equation (81) is

$$\frac{dJ}{dt} = \frac{e}{c} \oint \frac{ds'}{v_{\|}'} \mathbf{B}' \cdot \left\{ \sum_{j=2}^{n} \sum_{i=1}^{j-1} \left[\left(\frac{\partial \mathbf{r}}{\partial \alpha_i}\right)' \times \left(\frac{\partial \mathbf{r}}{\partial \alpha_j}\right)' \left(\dot{\alpha}_i'\dot{\alpha}_j - \dot{\alpha}_j'\dot{\alpha}_i\right) \right] \right\} \tag{85}$$

where $\dot{\alpha} = d\alpha/dt$. For every term in equation (85) the vector $\mathbf{S}_{ij} = (\partial \mathbf{r}/\partial \alpha_i) \times (\partial \mathbf{r}/\partial \alpha_j)$ represents the area of a parallelogram connecting four lines of force: the original line, those lying at unit change of α_i and α_j from the original one, and the line reached by the addition of both these changes. This area is not necessarily aligned perpendicular to $\mathbf{B}$, but the product $\mathbf{B} \cdot \mathbf{S}_{ij}$ is the magnetic flux through the area $\mathbf{S}_{ij}$ and is constant along the entire magnetic flux "tube".

Strictly speaking J itself is not in general a constant, but averaging over one bounce period of a particle (i.e., over $T = \int dt = \int ds/v_{\|}$) we have

$$\left\langle \frac{dJ}{dt} \right\rangle = \frac{e}{Tc} \oint \oint \frac{ds'ds}{v_{\|}'v_{\|}} \sum_{j=2}^{n} \sum_{i=1}^{j-1} \left\{ \mathbf{B} \cdot \mathbf{S}_{ij} \left[\dot{\alpha}_i(s')\dot{\alpha}_j(s) - \dot{\alpha}_j(s')\dot{\alpha}_i(s) \right] \right\} = 0. \tag{86}$$

Thus, as Northrup and Teller (1960) established, the average rate of change of J vanishes, because of an antisymmetry in changes in the rate of displacement of the line at s and at s'. The change in J due to ds' while the particle is on ds just cancels the change in J due to ds while the particle is on ds'. This cancellation applies to all pairs of line elements, ds and ds', and for all pairs of parameters, α_i and α_j, fixing the line.

2.5 Hamiltonian Representation: Adiabatic Invariants in Mechanical Systems

The two adiabatic invariants for charged particles in magnetic fields are special cases of adiabatic invariants of mechanical systems. Those systems that can be described by the ordinary Hamiltonian equations are treated by Landau and Lifshitz (1960); this discussion is a slight generalization of their presentation.

Hamilton's equations in terms of generalized position and momentum coordinates, q_i and p_i, are

$$\frac{dq_i}{dt} = \frac{\partial H}{\partial p_i}, \qquad \frac{dp_i}{dt} = -\frac{\partial H}{\partial q_i}, \tag{87}$$

where the total energy of a particle is given by the Hamiltonian,

$$H = \sum_j \frac{p_j^2}{2m} + U, \tag{88}$$

and U is the potential energy, which depends only on the q_i's. Ordinarily $p_i = mv_i$ is the particle momentum.

For a charged particle in a magnetic field, the usual form of the Hamiltonian must be altered. One might conceivably consider the interaction between a particle and magnetic field as providing a potential energy; but since it depends on the velocity of the particle, it is more conveniently regarded as providing kinetic energy (in one coordinate, while draining it in another).

A consistent representation is obtained by writing $U = e\Phi$, where Φ is the electrostatic potential, and re-defining the momentum variable as

$$p_i = mv_i + \frac{e}{c}A_i, \tag{89}$$

where **A** is the magnetic vector potential. The Hamiltonian still represents the total particle energy and is (Morse and Feshbach 1953, p. 294)

$$H = \frac{1}{2m}\sum_j \left(p_j - \frac{e}{c}A_j\right)^2 + e\Phi. \tag{90}$$

Application of the first relation (87) yields equation (89), while the second expression (87) gives the equation of motion (34), as we shall now show.

We have by differentiation

$$\begin{aligned}\frac{dp_i}{dt} &= \sum_j \frac{1}{m}\left(p_j - \frac{e}{c}A_j\right)\frac{e}{c}\frac{\partial A_j}{\partial q_i} - e\frac{\partial \Phi}{\partial q_i} \\ &= \sum_j \frac{e}{c}\frac{\partial}{\partial q_i}(v_j A_j) - e\frac{\partial \Phi}{\partial q_i}.\end{aligned} \tag{91}$$

Substituting equation (89) for p_i on the left and writing equation (91) in vector form gives

$$m\frac{d\mathbf{v}}{dt} = \frac{e}{c}\mathbf{v}\times(\nabla \times \mathbf{A}) + \frac{e}{c}\mathbf{v}\cdot\nabla A - \frac{e}{c}\frac{d\mathbf{A}}{dt} - e\nabla\Phi = e\left(\frac{\mathbf{v}\times\mathbf{B}}{c} + \mathbf{E}\right), \tag{92}$$

which is the equation of motion (34).

We shall now use the Hamiltonian formulation of the equation of motion to derive the transverse adiabatic invariant of a particle in an electric and magnetic field. Consider some parameter, α, which specifies a characteristic of the external fields, and suppose that α varies slowly over a period of oscillation. The particle, obeying the equations of motion, experiences a time rate of change

$$\frac{d\alpha}{dt} = \frac{\partial\alpha}{\partial t} + \mathbf{v}\,.\,\nabla\alpha.$$

The change in α may be associated with a time change in the total energy, due to changing **E** or **B**. Then the Hamiltonian changes at a rate

$$\frac{dH}{dt} = \frac{dH}{d\alpha}\frac{d\alpha}{dt} = \frac{\partial H}{\partial\alpha}\frac{\partial\alpha}{\partial t}. \tag{93}$$

The action integral,

$$S = \oint p_i dq_i, \tag{94}$$

changes at a rate

$$\frac{dS}{dt} = \oint dq_i\left[\frac{\partial p_i}{\partial\alpha}\frac{d\alpha}{dt} + \frac{\partial p_i}{\partial H}\frac{dH}{dt}\right], \tag{95}$$

where the integral is to be evaluated at some fixed instant of time. The integration path is not in general the actual path of a particle; thus at a fixed instant of time, the Hamiltonian is constant over the path, or $dH/d\alpha = 0$.

For a circular path, the velocities, dq_j/dt, in any direction are the same for any point on the path, q_i, as they are at the particle. The first term in equation (95) is then equivalent to dp_i/dt for an identical particle at any point on the path and we have, by equation (87),

$$\frac{dS}{dt} = -\oint dq_i\frac{\partial H}{\partial q_i} = 0. \tag{96}$$

Now let us consider the application of equations (94) and (96) to μ. Writing $p = p_\varphi = -p_\perp$ and $dq = \rho d\varphi$, and eliminating ρ with equation (33), we find from equations (89) and (94)

$$S_\perp = -\oint\frac{mv_\perp{}^2}{\omega}d\varphi + \frac{e}{c}\int\mathbf{B}\cdot d\boldsymbol{\sigma}, \tag{97}$$

where the second integral, obtained with the aid of Stokes' theorem, extends over the area, $\sigma = \pi\rho^2$, of a gyration. From equation (69) we may then write

$$S_\perp = -\frac{2\pi mc}{e}\mu = \text{const.} \tag{98}$$

Thus the constancy of μ, the transverse invariant, is obtained as a special case of equation (96).

When there are no external forces parallel to the path, the generalized momentum, p_i, as defined by (89) is constant. In this special case the invariance of μ follows directly from the condition that $dp_i/dt = 0$, and it may therefore be regarded as a consequence of the conservation of generalized momentum.

It is a general characteristic of classical oscillators that H/ω = const, where H is the total energy and ω is the frequency of oscillation. The gyrating particle in a purely time-dependent magnetic field is a special case of a classical oscillator.

On this general basis the constancy of J was first hypothesized by M. Rosenbluth. However, the derivation is more complex than it is for μ. For gyrations about the field, the motion in a curved path is included in the formulation of the Hamiltonian. But for motions of the guiding center along a curved path, the centrifugal force must be inserted, either in the fashion of equation (76) or by modifying the Hamiltonian for a rotating coordinate system. Moreover the velocities are not constant along the integration path, q_i, and therefore are not equivalent to the corresponding terms in the Hamiltonian equations of motion of the particle at the same q_i. For this reason the derivation of the constancy of J is rather elaborate and requires a demonstration of the mutual-cancellation effect, as discussed in the previous section.

2.6 Coordinate Systems for Particles in the Geomagnetic Field

The Störmer problem discussed earlier involves a detailed calculation of an orbit of a particle through the geomagnetic field (approximated as a dipole). Hence ordinary space coordinates (r, λ, φ or R, z, φ) are appropriate. One might even allow for some unsymmetrical features of the earth's field by approximating it with an off-center dipole.

The guiding-center approximation used in the description of trapped plasmas lends itself to other coordinate systems than purely spatial ones. One is not usually interested in the instantaneous location of a specific particle, but only in the principal features of the orbits.

Acceleration mechanisms and dumping (bombardment into the atmosphere) may be discussed in terms of the magnetic moment, μ, and the kinetic energy, ϵ. Particle drifts are described by μ and J.

Particular interest is attached to the particle motions when ϵ does not vary with time. In general, the individual particles, with different mirror points, on a line of force will drift to various different lines of force. They all return to the original one, by virtue of the constancy of I defined by equation (73). But at intermediate lines of force, a thin shell is broadened. This dispersion is small for the earth's field, however, and all particles starting from one line of force will remain in a fairly thin shell.

Once a shell is specified, B = constant will denote two lines (one in each hemisphere) on the shell, and these lines will trace mirror points for a homogeneous group of particles. The problem is to find some general parameter, L, which will substitute for the equatorial radius, r_0, of a shell of lines of force for a dipole. If L is any function of only I and B, it will have the correct longitude dependence, since $I(B_m)$ = constant specifies the lines of force swept over by drifting particles with mirrors at B_m = constant. Indeed, I for a definite value of B_m would itself uniquely define a shell. However it is not especially convenient; it is analogous to defining a shell in a dipole field in terms of the arc lengths outside a surface of specified field strength.

McIlwain (1961) has obtained a parameter L, analogous to r_0, which is widely used for mapping trapped-particle belts. For a dipole, once a shell r_0 is specified, the latitude $\lambda = \lambda_m$ designates a definite arc length—i.e., a definite I. Thus

$$I = r_0 f_1(\lambda).$$

Also $B = M r_0^{-3} f_2(\lambda)$ by equation (6). Hence

$$\frac{I^3 B}{M} = \frac{r_0^3 B}{M} f_1^3(\lambda) = \frac{r_0^3 B}{M} f_3\left(\frac{r_0^3 B}{M}\right) = f_4\left(\frac{r_0^3 B}{M}\right) \tag{99}$$

or

$$\frac{r_0^3 B}{M} = F\left(\frac{I^3 B}{M}\right). \tag{100}$$

McIlwain has computed the function F for a dipole field and then defined L by

$$\frac{L^3 B}{M} = F\left(\frac{I^3 B}{M}\right), \tag{101}$$

where I and B now pertain to a point in the real field. Thus one obtains I and B for a point in the real field, computes F and then L. In this way the two parameters L and B are sufficient to characterize a shell and any mirror line on that shell; this information is sufficient for most mapping purposes.

3. Hydromagnetic Theory of Drift Motions

The guiding-center theory treated in Section 2 is useful not only for describing the motions of energetic (auroral or Van Allen) particles trapped in the geomagnetic field, but is also applicable to the "background" plasma with thermal energies, which constitutes the vast majority of trapped particles. The theory, treating individual particles, is inherently more rigorous than the macroscopic fluid theory, and has even been applied to the analysis of certain instabilities (Rosenbluth and Longmire 1957).

For many purposes, however, it is far more convenient to describe a plasma with the macroscopic equations, even when collisions between particles are negligible, as they are in most problems involving geomagnetically trapped particles. The hydromagnetic theory is well developed and, when appropriate to the situation at hand, is highly useful for discussions of flow of the plasma, hydromagnetic waves, and instabilities.

However the application of these equations is not always appropriate. Because of the strong control of the geomagnetic field, intuitive analogies with pure hydrodynamic flow are hazardous. For example, the use of hydrodynamic analogies to describe the circulation of geomagnetically trapped plasma should always be justified by more rigorous means, since the driving forces operate indirectly through interactions of the plasma with the electromagnetic fields (including those on the boundaries) and not directly through the fluid interactions with itself.

3.1 Basic Equations

The fundamental macroscopic equations may be derived from the Boltzmann equation (Spitzer 1956; Chandrasekhar, Kaufman, and Watson 1957). Let $\mathbf{V}$ be the mean mass velocity, ρ the mass density, P the pressure (assumed to be isotropic), $\mathbf{J}$ the current density (in electromagnetic units), σ the ordinary conductivity, and $\mathbf{f}$ an acceleration due to an external force. The linearized (i.e., with quadratic terms in $\mathbf{V}$ and $\mathbf{J}$ and their derivatives discarded) equation of motion for the case of electrical neutrality ($N_e = N_i$) is then

$$\rho\frac{\partial \mathbf{V}}{\partial t} = -\nabla P + \rho\left(\mathbf{f}_i + \frac{m_e}{m_i}\mathbf{f}_e\right) + \mathbf{J} \times \mathbf{B}. \tag{102}$$

Here subscripts e refer to electrons; i will denote positive ions. The generalized Ohm's law in a steady state (i.e., with an inertial term, $\partial \mathbf{J}/\partial t$, discarded) is

$$\mathbf{E} + \frac{\mathbf{V} \times \mathbf{B}}{c} = \frac{c}{\sigma}\mathbf{J} + \frac{1}{N_e e}[\mathbf{J} \times \mathbf{B} + N_e m_e \mathbf{f}_e - \nabla P_e]. \tag{103}$$

For examining the general behavior of a fluid, these equations must be solved simultaneously with Maxwell's equations, the equation of continuity, and an equation of state (relating P and N). (See, for example, Chandrasekhar, Kaufman, and Watson 1957; Kruskal 1960a).

3.2 Drift Motions

Here we shall be concerned only with the steady-state flow of a hydromagnetic fluid perpendicular to the magnetic field. This is the same problem treated in Section 2.2 with the guiding-center approach. If we consider the external force to be the centrifugal force arising from motion along the field lines, $\mathbf{f}$ is always perpendicular to $\mathbf{B}$. By equation (102), the steady-state pressure gradient then vanishes parallel to $\mathbf{B}$. Similarly from equation (103) with an infinite conductivity (which is roughly appropriate for a highly ionized gas), $\mathbf{E}$ is zero along the field.

Substituting equation (102) into (103) gives

$$\mathbf{E} + \frac{\mathbf{V} \times \mathbf{B}}{c} = \frac{1}{N_i e}(\nabla P_i - \rho \mathbf{f}_i). \tag{104}$$

The vector product of $\mathbf{B}$ times this equation gives $V_{||} = 0$ and a perpendicular "drift" velocity,

$$\mathbf{V} = \frac{c}{e}\frac{\mathbf{B}\times}{B^2}\left(-e\mathbf{E} + \frac{\nabla P_i}{N_i} - m_i \mathbf{f}_i\right). \tag{105}$$

To relate this formula to that obtained from the particle-drift theory, let us make the simplifying assumption of a uniform particle density and take an equation of state

$$P_i = \frac{N_i}{2} m_i \langle v_{i\perp}{}^2 \rangle, \tag{106}$$

where $\langle v_{i\perp}{}^2 \rangle$ is the mean-square gyrational velocity. (In an isotropic medium, the transverse energy components contain 2/3 the mean ion energy of (3/2) kT_i.) If the magnetic moments of the individual particles are conserved during drifts, we have $m_i \langle v_{i\perp}{}^2 \rangle /2 = \langle \mu \rangle B$, where $\langle \mu \rangle$ is constant, and

$$\nabla P_i = \frac{N_i m_i \langle v_{i\perp}{}^2 \rangle \nabla B}{2B}. \tag{107}$$

Equation (105) is then identical to equation (80) averaged to represent the mean drift velocity of the ions.

Similarly, the vector product of **B** with equation (102) yields

$$\mathbf{J} = \frac{\mathbf{B} \times}{B^2} [\nabla(P_i + P_e) - N_i m_i \mathbf{f}_i - N_e m_e \mathbf{f}_e], \tag{108}$$

which is $N_e(e/c)(\mathbf{V}_i - \mathbf{V}_e)$.

Part of the difficulty in applying the hydromagnetic equations to low-density plasmas is merely conceptual. In a high density fluid the individual molecules might diffuse in and out of a moving volume rather slowly, so that the volume retains some physical identity for long periods. The macroscopic equations govern its motion as a whole. For the plasma around the earth, the mass velocity **V** is formally defined as

$$\mathbf{V} = \frac{\sum m_\alpha v_\alpha}{\sum m_\alpha} \tag{109}$$

where the summations are carried over all particles in unit volume. While the hydromagnetic equations still govern **V**, they do not imply that a volume moves as a single entity.

Hence for drift motions, the macroscopic approach gives a mass motion the same as the mean drift velocity for the heavy ions at the same position, since

$$m_i \gg m_e.$$

There is thus no difference between the circulation of matter in the earth's magnetic field in the hydromagnetic (or convective) picture and the *mean* drift motions obtained with the guiding-center theory. In either case departures from simple "ring-currents" (in the trapped-particle sense) require driving electric fields.

A related point concerns motions of the magnetic lines of force. For an infinitely conducting fluid the material is often said to be "frozen" to the field lines in the sense that both move with the same transverse velocity. For a plasma one must bear in mind that formally the individual lines of force are not defined in electromagnetic theory; all the above statement implies is that the magnetic flux through a specified area remains constant as the area moves with the mean plasma velocity.

The total rate of change of flux through an element of area dS is

$$\int \frac{d\mathbf{B}}{dt} \cdot d\mathbf{S} = \int \left(\frac{\partial \mathbf{B}}{\partial t} + \mathbf{V} \cdot \nabla \mathbf{B} \right) \cdot d\mathbf{S} = - \int \nabla \times (c\mathbf{E} + \mathbf{V} \times \mathbf{B}) \cdot d\mathbf{S}, \quad (110)$$

where $\partial \mathbf{B}/\partial t$ has been substituted with $c\nabla \times \mathbf{E}$ in accordance with Faraday's law. Taking the curl of equation (103) for infinite conductivity and using (102) to eliminate $\mathbf{J} \times \mathbf{B}$ gives

$$\nabla \times (c\mathbf{E} + \mathbf{V} \times \mathbf{B}) = \nabla \times \left[\frac{c}{N_e e} (\nabla P_i - \rho \mathbf{f}_i) \right]. \quad (111)$$

In the special case where $\mathbf{f}_i$ may be expressed as the gradient of a scalar potential and when N_e is constant, equations (111) and (110) are zero. It is not, however, generally true that the mean motions of the plasma correspond to motions of the field lines (Spitzer 1956). Even when it is true, one must bear in mind that the plasma in the geomagnetic field extends in energy over the entire range from below one electron volt (thermal) to over a hundred MeV (Van Allen protons). The average behavior may be quite different from that of any particular interval in this spectrum.

In the orbit theory of Section 2, there is a third adiabatic invariant, the conservation of magnetic flux, which we have not considered. It is equivalent, however, to the conservation of flux with the macroscopic picture treated here.

References

Alfvén, H., 1950, *Cosmical Electrodynamics* (London and New York: Oxford University Press).

Chamberlain, J. W., 1958, Theories of the aurora, *Advances in Geophysics*, ed. H. E. Landsberg and J. Van Mieghem (New York: Academic Press), pp. 109–215.

—— 1961, *Physics of the Aurora and Airglow* (New York: Academic Press), Chapter 3.2.4, pp. 77–79.

Chandrasekhar, S., Kaufman, A. N., and Watson, K. M., 1957, Properties of an ionized gas of low density in a magnetic field, III, *Ann. Phys.*, **2**, 435–470.

Kruskal, M., 1960a, Hydromagnetics and the theory of plasma in a strong magnetic field, and the energy principles for equilibrium and for stability, *The Theory of Neutral and Ionized Gases*, ed. C. DeWitt and J. F. Detoeuf (New York: J. Wiley & Sons), pp. 253–274.

—— 1960b, Asymptotic theory of systems of ordinary differential equations with all solutions nearly periodic, *The Theory of Neutral and Ionized Gases*, ed. C. DeWitt and J. F. Detoeuf (New York: J. Wiley & Sons), pp. 277–284.

Landau, L., and Lifshitz, E., 1951, *The Classical Theory of Fields* (Cambridge, Mass: Addison-Wesley).

—— 1960, *Mechanics* (London: Pergamon Press).

McIlwain, C. E., 1961, Coordinates for mapping the distribution of magnetically trapped particles, *J. Geophys. Res.*, **66**, 3681–3691.

Morse, P. M., and Feshbach, H., 1953, *Methods of Theoretical Physics* (New York: McGraw-Hill).

Northrup, T. G., 1961, The guiding center approximation to charged particle motion, *Ann. Phys.*. **15**, 79–101.

Northrup, T. G., and Teller, E., 1960, Stability of the adiabatic motion of charged particles in the earth's field, *Phys. Rev.*, **117**, 215–225.

Poincaré, H., 1896, Remarques sur une expérience de M. Birkeland, *Compt. rend.*, **123**, 530–533.
Rosenbluth, M. N., and Longmire, C. L., 1957, Stability of plasmas confined by magnetic fields *Ann. Phys.*, **1**, 120–140.
Spitzer, L., 1956, *Physics of Fully Ionized Gases* (New York: Interscience).
Störmer, C., 1907, Sur les trajectoires des corpuscles électrisés dans l'espace sous l'action du magnétisme terrestre avec application aux aurores boréales, *Arch. sci. phys. et nat.*, [4] Genève, **24**, 5–18, 113–158, 221–247, 317–354.
—— 1911, Sur les trajectoires des corpuscles électrisés dans l'espace sous l'action du magnétisme terrestre avec application aux aurores boréales, *Arch. sci. phys. et nat.*, [4] Genève, **32**, 117–123, 190–219, 277–314, 415–436, 501–509; 1912, *Ibid.*, **33**, 15–69, 113–150.
—— 1955, *The Polar Aurora* (London and New York: Oxford University Press).
Vestine, E. H., and Sibley, W. L., 1959, Remarks on auroral isochasms, *J. Geophys. Res.*, **64**, 1338–1339.

INDEX